The Case the Missing Cookies

by Thomas K. and Heather Adamson • Illustrated by Charlie Alder

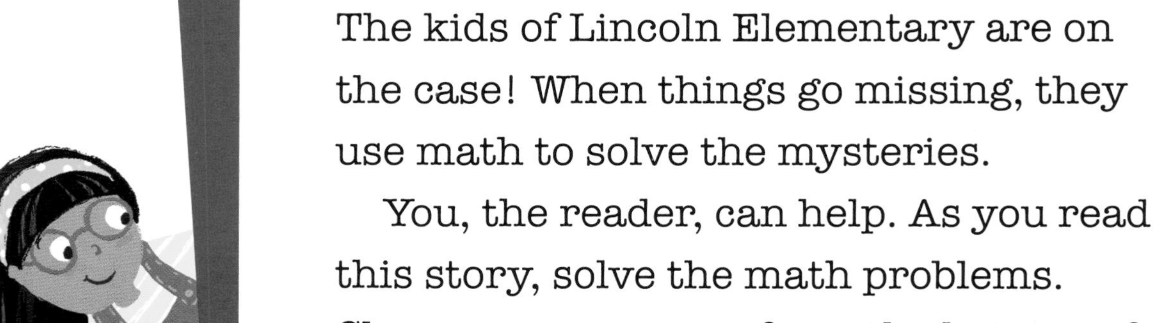

The kids of Lincoln Elementary are on the case! When things go missing, they use math to solve the mysteries.

You, the reader, can help. As you read this story, solve the math problems. Choose your answer from the bottom of the page, and skip to the page for that answer. If you're right, you're one clue closer to solving the mystery! If you're wrong, you can try again.

What do YOU think happened to the missing cookies?

Start on page 3.

Hurray! Mr. Jeb's class won this month's reading prize. The class gets two boxes of cookies from Mrs. Bonn's bakery. Each box has 12 cookies.

Gina knows there are 19 kids in her class. She wonders, are there enough cookies for the whole class?

If there are enough cookies, turn to page 6. ▸

If there are not enough cookies, turn to page 8. ▸

Think again.

Add the weekly totals together.

Miss Krump's class was a close second!

Turn to page 10. ▸

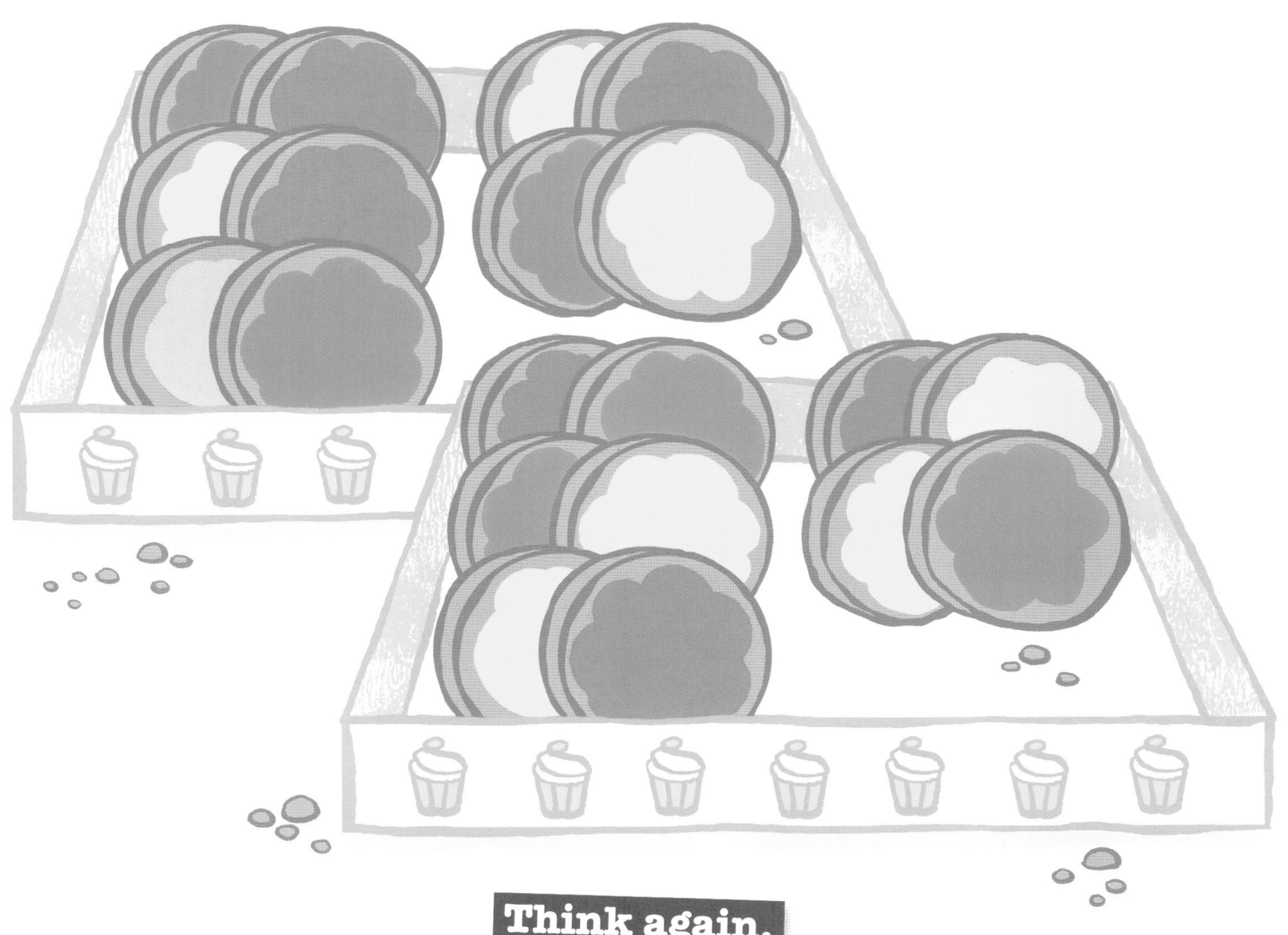

Think again.

Skip counting is counting in equal groups.

Gina skip counted by 2s, and she counted 10 pairs.

2, 4, 6, 8, 10, 12, 14, 16, 18, 20

Turn to page 12. ▸

There are more than enough cookies!

$$12 + 12 = 24$$

And 24 is more than 19. Gina can't wait to have a cookie after lunch.

Go to page 7.

The class returns to the classroom, ready for their treat. Gina notices that the boxes have been opened. "Mr. Jeb, there are cookies missing!" she says.

Go to page 9.

Think again.

With 2 boxes of 12 cookies, Gina can add 12 + 12.
That equals 24. That's enough for the 19 kids in the class.

◂ **Turn to page 6.**

"How can you find out for sure?" Mr. Jeb asks her.

Gina knows there was an even number, so she skip counts by 2s. How many cookies are left?

If 20 cookies are left, turn to page 12. ▸

◂ If 22 cookies are left, turn to page 5.

“It was Miss Krump’s class!” says Gina.

“It was close, too,” says Jaden. “Maybe they weren’t happy about it.”

“We don’t know for sure if that’s what happened,” says Mr. Jeb. “Let’s use our detective skills to look for more clues.”

Gina suggests, “We could group the cookies by color.” Which color is missing?

If yellow is the missing color, turn to page 15. ▸

If blue is the missing color, turn to page 18. ▸

Yes, Gina counted 20 cookies.

"Now we can use subtraction to find out how many fewer cookies there are," Mr. Jeb says.

"It's easier to see on the number line," Gina adds. "Then we'll know how many are missing."

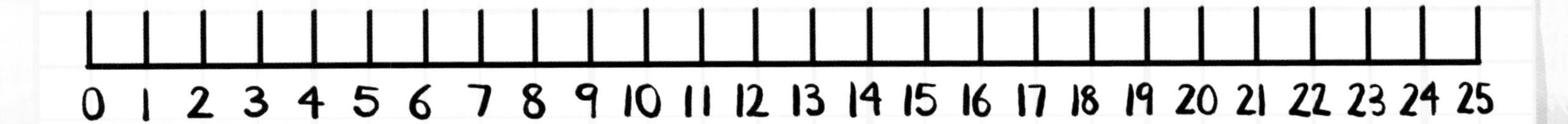

Are 2 cookies missing? Turn to page 14. ▸

Are 4 cookies missing? Turn to page 16. ▸

Think again.

Count the spaces on the number line.

You can count 4 spaces from 24 to 20.

$$24 - 20 = 4$$

So four cookies are missing.

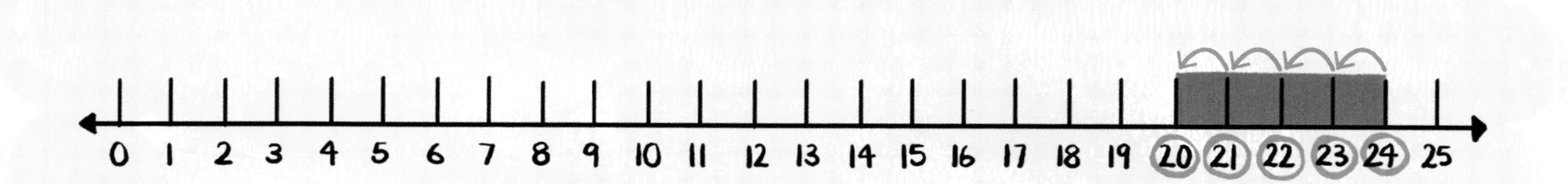

Go to page 16.

Think again.

When all the cookies are arranged in rows by color, which color has the least?

Go to page 18.

Yes, four cookies are missing.

Gina wants to find more clues.

"Who would have taken four cookies?" Jaden asks.

"Hey, what class came in second in the reading contest? Would they do this?" wonders Camila.

Gina says, "Our class read 57 books."

Here are the totals for the other classes. Which class came in second?

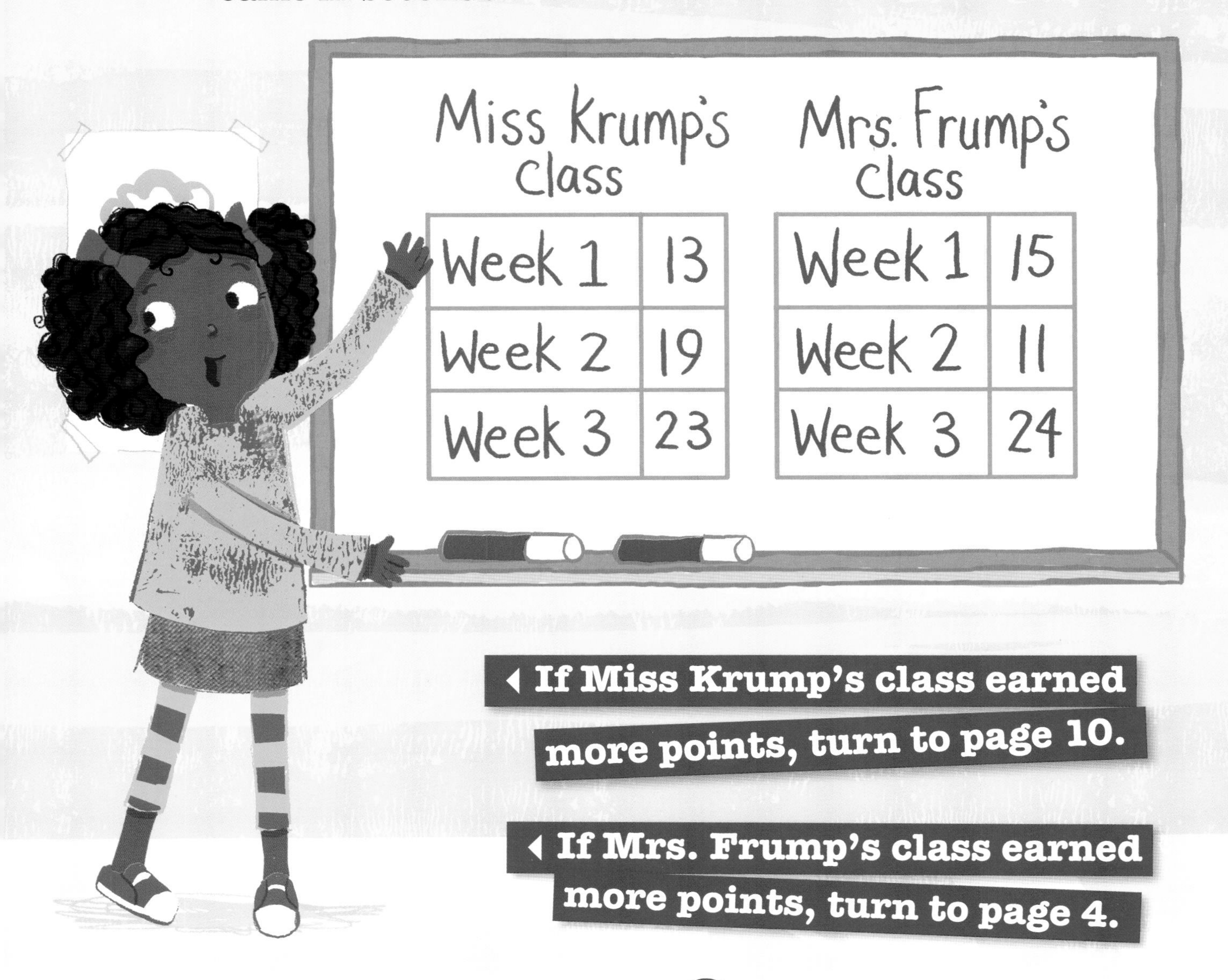

If Miss Krump's class earned more points, turn to page 10.

If Mrs. Frump's class earned more points, turn to page 4.

"Blue has the fewest cookies!" says Gina. "Wow, someone took only blue ones."

In the office, Gina tells the principal, "Mrs. Penn, we want to report missing cookies."

Mrs. Penn says, "How did you know some were missing?"

"We used math clues. We counted by 2s and there were four cookies missing after lunch," Gina answers.

"You still have enough for your class, don't you?" asks Mrs. Penn. "There are 10 boys and 9 girls in your class."

If there are not enough cookies, turn to page 20.

If there are still enough cookies, turn to page 21.

Think again.

Ten boys and nine girls adds up to 19 students. That still leaves 20 cookies for 19 kids. But what happened to the missing cookies?

Go to page 21.

Gina says, “Yes, we still have 20 cookies left. 20 is more than 19.”

Mrs. Penn says, “So you still have enough cookies for your class, even if a few blue ones are missing.”

Gina says, “Mrs. Penn, I didn’t say it was blue ones that were missing.”

Go to page 22.

"You didn't? Oh," says Mrs. Penn.

Gina sees Mrs. Penn's face. There's a bit of blue frosting by her mouth!

"Um, Mrs. Penn, did you eat four blue cookies?"

"Blue is my favorite color!" Mrs. Penn exclaims. "I ate three cookies, and saved one for Mr. Jeb."

"Besides, I knew there were extra cookies. I did the math. There are still enough cookies for all the students," says Mrs. Penn.

This mini math mystery is solved!

AMICUS ILLUSTRATED and **AMICUS INK**
are published by Amicus
P.O. Box 227, Mankato, MN 56002
www.amicuspublishing.us

Library of Congress Cataloging-in-Publication Data
Names: Adamson, Thomas K., 1970- author. | Adamson, Heather, 1974- author. | Alder, Charlie, illustrator.
Title: The case of the missing cookies / by Thomas K. and Heather Adamson ; illustrated by Charlie Alder.
Description: Mankato, Minnesota : Amicus, [2022]. | Series: Mini math mysteries | Audience: Ages 6-9. | Audience: Grades 2-3. | Summary: "Who took the cookies? Be a math detective along with the kids at Lincoln Elementary in this pick-your-own-path mystery. Be a detective and use your addition, subtraction, and skip counting skills to solve the case"—Provided by publisher.
Identifiers: LCCN 2019045889 (print) | LCCN 2019045890 (ebook) | ISBN 9781645490104 (library binding) | ISBN 9781681526522 (paperback) | ISBN 9781645490906 (pdf)
Subjects: LCSH: Arithmetic—Juvenile literature. | Mathematical recreations—Juvenile literature. | Cookies—Juvenile literature.
Classification: LCC QA115 .A3325 2022 (print) | LCC QA115 (ebook) | DDC 513.2—dc23
LC record available at https://lccn.loc.gov/2019045889
LC ebook record available at https://lccn.loc.gov/2019045890

Editor: Rebecca Glaser
Designer: Kathleen Petelinsek

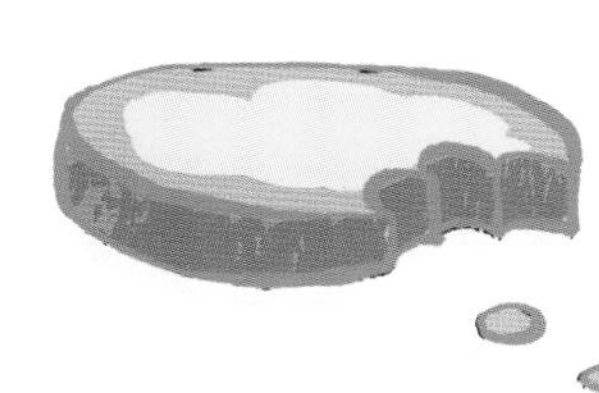

About the Authors

Thomas K. and Heather Adamson have both written many books for kids. This husband-and-wife team enjoys writing together. When they are not working, the couple likes to take hikes, watch movies, eat pizza, and of course, read. They live in South Dakota with their two sons and a Morkie named Moe.

About the Illustrator

Charlie Alder has written and illustrated many books for children, including *Daredevil Duck* and *Chicken Break!* She lives in Devon, England, with her husband and son. When not drawing chickens, Charlie can be found in her studio drinking coffee, arranging her crayons, and inventing more cheeky accidental heroes, both human and animal.